A horse engine with a one-up, one-down arrangement shown on the Low Hall Colliery, Cumberland, token of 1797.

ANIMAL-POWERED MACHINES

J. Kenneth Major

Shire Publications Ltd

CONTENTS

Set in 9 point Times roman and printed in Great Britain by C. I. Thomas & Sons (Haverfordwest) Ltd, Press Buildings, Merlins Bridge, Haverfordwest, Dyfed.

British Library Cataloguing in Publication Data
Major, J. Kenneth
  Animal-powered machines.—(Shire album; v.128)
  1. Animal-powered engines—History
  I. Title
  621.4   TJ830
  ISBN 0-85263-710-1

ACKNOWLEDGEMENTS
  The author wishes to thank all those who have helped in the preparation of this book and all those who have allowed him access to their animal-powered machines so that these could be studied fully. Particular thanks are given to the members of the Wind and Watermill Section of the Society for the Protection of Ancient Buildings and the International Molinological Society, to the Museum of English Rural Life at Reading, the Museum of London and the Welsh Folk Museum at St Fagans. Thanks are also due to Hugo Brunner, Martin Watts and R. Shorland-Ball, and to Helen Major for her typing and support in this venture. Photographs on the following pages are acknowledged to: Miss Gardner, pages 23 (upper), 24, 26 (centre); the Museum of English Rural Life, 25, 26 (upper); the Tankerness House Museum, 26 (upper).

COVER: *The Donkey Wheel at Broad Hinton, Wiltshire, demolished in 1908. From a coloured postcard.*

A horse-driven pump at Tidworth in Wiltshire at work during the 1930s.

The donkey wheel at Carisbrooke Castle on the Isle of Wight. A lithograph of the 1850s.

INTRODUCTION

Animal-powered machines are those in which the effort of animals is converted to rotary power; the term does not relate to the movement of vehicles in which a horse, for example, provides tractive effort. There are many examples of prime movers in Britain, that is machines powered by wind or water, but there are very few examples of animal-powered machines, although there is now much of evidence to show that they were widely in use in the period just before the universal use of the steam engine.

It is convenient to divide animal-powered machines into various classifications. The most obvious subdivisions of the types are the *vertical* and the *horizontal*. This classification follows that which is normally ascribed to the waterwheel. The vertical machines are those in which the animal or man moves in a vertical plane, and this motion gives rotary motion to a horizontal shaft. The horizontal machine is one in which the animal moves in a circle about a central pivot, and the rotary motion can be transmitted by gearing from a gear ring formed around the same central pivot. In addition to these two basic types there are machines which cannot be placed in either category, that is where animals move in an *oblique* plane to give rotary

3

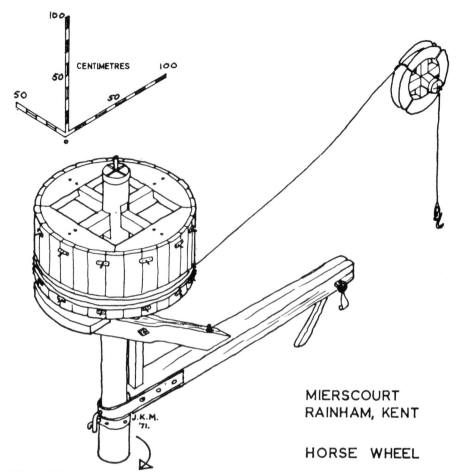

MIERSCOURT
RAINHAM, KENT

HORSE WHEEL

The single-horse water-raising wheel at Mierscourt near Rainham in Kent. When the horse stops the sprag prevents the rope and bucket from running back.

motion by means of belts or gears.

The vertical machine is one in which the animal, or man, treads a series of boards on the circumference of a circle and this rim transmits the motion to the centre of the wheel and turns the shaft by means of its spokes. The animal or man can be on the inside of the rim, in which case the machine is known as a *donkey wheel* or *treadwheel*, or a man can work on the outside of the circumference and this is then known as a *treadmill*.

The horizontal machines have further subdivisions and these can be classified either as *direct*, where the power is performing its function without the presence of gearing, or as *indirect*, where the power functions only by the application of gearing between the motion of the animal and the work the machine does. The direct examples fall into two further categories. The first machines are the *roller crushers*, and in these an animal drags an edge-runner stone around in a trough to crush fruit or oil seed to obtain juices or fluids. The second category covers the *rope-winding machines*, often called *horse gins*, and in these a winding drum is mounted around a vertical shaft which is turned by one or more horses

ABOVE: *An apple crusher from Pyne's 'Microcosm' of 1803.*
BELOW: *The horse wheel for water pumping at Earlham Hall, Norwich. This was worked by a single horse and is a typical wooden-geared horse engine.*

harnessed to arms below the winding drum.

The indirect examples also fall into two further categories. One can be called the *horse engine* and it was usually used for agricultural purposes. This is a low-level gear, usually of cast iron and quite small in diameter, with arms to which the horses are harnessed above the gear ring. The gear is connected by means of bevel gears to a lay shaft which runs to machinery placed outside the circular path traversed by the horses. The second category can conveniently be called a *horse wheel.* In this a large-diameter wheel, usually made of wood, is mounted on a vertical shaft at high level and the horses are harnessed to horns bolted to the spokes of the wheel. This large-diameter wheel can be fitted either with wooden gear teeth or have a cast iron gear ring bolted above it. The lay shaft taking the power from this wheel by means of a bevel gear usually runs above the horse wheel to an adjacent barn, in which it would drive fixed farm machinery.

The animal-powered machines outside these categories are the oblique treadmill and the oblique treadwheel. The *oblique treadmill* consists of a moving belt set at an angle to the ground, on which a dog or horse is tethered. As the animal walks, the belt moves away behind it and thus gives motion to a horizontal shaft by means of pulleys mounted on it which rotate with the belt. The *oblique treadwheel* consists of a circular treadboard mounted on a shaft which is set at an angle. As the animal treads the board, it falls away beneath it giving rotary motion to the shaft. In Britain, the only animals known to have worked these are dogs. Usually gearing mounted under the treadboard gives rotation to a lay shaft by means of bevel wheels.

These animal-powered machines can be seen in many places and can form a fascinating study as more and more are being discovered. The surviving examples give only a small picture of the many purposes for which these machines were used.

A photograph of around 1900 of an oblique tread-mill in use on a farm in southern England. The horse is harnessed above the belt which moves away from under its hooves.

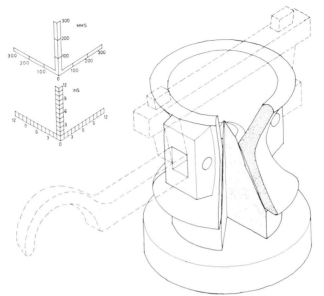

The hourglass mill found in Princes Street, London, which dates from the first or second century AD. This had been imported from the Eifel area of Germany and is made of basalt. The arms are conjectural and would be suitable for a donkey to work.

ANIMAL-POWERED MACHINES
IN HISTORICAL TIMES

The Romans depicted two important machines in their bas-reliefs: these are the treadwheel crane and the 'hourglass' corn mill. The treadwheel crane has not been recorded in Roman Britain but the hourglass corn mill has. The obliteration of Pompeii by lava sealed in four hourglass mills, which have been recorded by excavation of the baker's shop. In London, a similar hourglass mill was found in Princes Street, and this has been re-erected in the Museum of London. This hourglass mill is cut out of a single piece of Eifel basalt from the lava field west of the Rhine between Koblenz and Bonn. It consists of two cones which meet to form the hourglass shape. The diameter at top and bottom is 27 inches (685 mm) and the diameter at the waist is 16 inches (405 mm). The height is 23 inches (585 mm). The projections at the sides carried the arms to which the donkey was harnessed. This direct-drive machine, in which the animal was directly connected to the moving parts by the horse arm, had several parallels. The Romans crushed olives in a circular trough in which edge-runner stones rotated, and from this form the fruit crusher, such as the edge-runner cider mill, developed.

Water raising was an essential service in prehistoric and Roman times although it was not really much required in Britain. In the commonest examples the water was raised in a chain of buckets or pots powered by an animal wheel at ground level. This was developed in parallel with the vertical treadwheel, such as the donkey wheel, in which the power of the animal treading a wheel of large diameter raised water by winding up a bucket rope on a drum on the main shaft. These machines gradually came to be used in large houses towards the end of the medieval period, whilst the chain of buckets was used increasingly for the dewatering of mines. There are sixteenth-century examples of the vertic-

ABOVE: *The Brixton treadmill from an illustrated paper of about 1830. The treadmill is shown without the partitions which came to be used following the introduction of the Separate System in the 1840s.*

BELOW: *The treadmill at Beaumaris gaol on Anglesey was originally built with cast-iron separating walls between the bays but was subsequently altered with a wooden partition to house another prisoner. It is shown connected to both the pumps and a brake.*

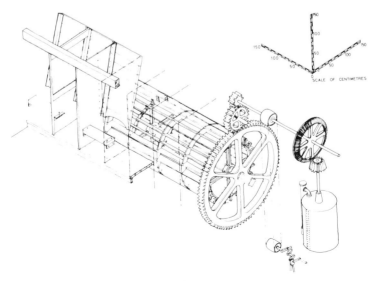

A dog-driven spit in the kitchen of the manor house of St Fagans, the Welsh Folk Museum.

al wheel in existence in Britain, and it may well be that they were fairly widely distributed in the medieval period in the chalk upland areas of south-east England, where wells have to be deep.

There are vertical wheels in some cathedrals, used for hoisting materials above the vaults of the nave and choir and for hoisting timbers to the roof and the tower. This type is a precursor of the punitive treadmill. The use of the treadmill as a means of punishment was re-invented by Sir William Cubitt in 1818 and came to be accepted in most prisons. The treadmill is a wheel on the outside of which there is a series of radially mounted steps on to which the prisoner treads and has to keep stepping as it runs away underneath him. The surviving example in Beaumaris has a tread of 9 inches (120 mm) and the diameter of the wheel is 5 feet (1.5 m). Six prisoners walked on the treadmill at a time and pumped water from a well to tanks in the roof. The machines appear to have been standardised, with the iron parts supplied from a central source. Whilst many of the treadmills were made to raise

water or work against a brake, others ground corn. At the time of the discontinuation of the use of the treadmill (about 1900), there were nineteen prison treadmills grinding corn for their own use and for other prisons. The treadmill at Reading was installed in 1828 and removed in the 1850s. After six months of use the warders ceased to be the millers at Reading and the power of the treadmill was leased to a local miller. He supplied the prison with its requirements and sold the finer meal in the town.

One element of animal power which dates back to the medieval period in Britain is the animal-powered spit. A wooden drum was usually mounted at high level on the wall over the open fireplace, and in this a dog trod the inside rim of the drum. On the main shaft of the drum there was a pulley wheel which worked another one on the spit and thereby turned the meat over in front of the fire. The pulley cord turned at right angles over pulleys, the dog treadwheel usually being parallel with the fireplace wall. A type of dog, the

9

An oblique dog-driven wheel from Dulellog, Rhostryfan, Caernarfon, erected at St Fagans. This drove the paddle in a churn by means of the gearing under the tread boards.

pedigree name of which was the Turnspit, was bred for this duty and looked like a wire-haired dachshund with a square muzzle. Thomas Bewick illustrates this dog in his *Quadrupeds*.

The natural progression from the dog spit was to the dog-driven churn. In this the dog provided the motion to drive the churn by means of pulleys and gears so that it rotated or, by means of cranks, was operated by dashing a plunger up and down in the milk. Many examples of these exist in museums outside the United Kingdom and some later examples are displayed at the Welsh Folk Museum at St Fagans. Oblique treadwheels and paddle engines were also used to provide the power for butter making. These examples are important and their use should be recorded when they are discovered.

It is understandable that so few examples of animal-powered machines which date from before the industrial revolution exist in Britain. Those which remain form a unique part of technical history and should be cherished.

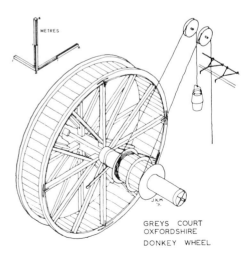

GREYS COURT
OXFORDSHIRE
DONKEY WHEEL

The donkey wheel at Greys Court in Oxfordshire.

10

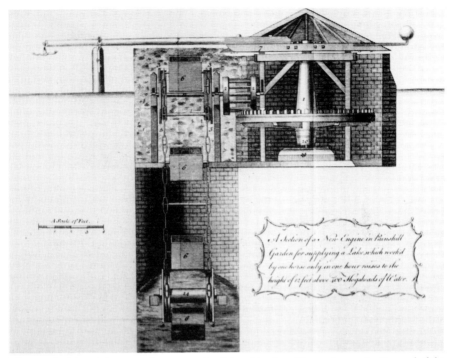

A horse wheel for raising water by means of a chain of buckets. The engine carries the roof of the building round with it, and was in use at Painshill, Surrey, from 1770 until the 1830s.

VERTICAL HOISTING MACHINES

The principal use of animal-powered machines that move in a vertical plane was in the hoisting of materials and water. The machine consists of a large-diameter wheel in which men or animals tread the inside of the rim of the wheel. This rim is usually attached to the shaft by two sets of spokes which are set either compass-armed or clasp-armed about the shaft. Compass-armed spokes are mortised through the shaft and are halved and wedged to each other so that they sit in one plane and are tight in the mortise. Clasp-armed spokes are framed and halved together in such a way that a square hole is left through which the shaft is threaded and fixed by means of wedges. The compass-armed wheel is weaker than the clasp-armed wheel because it reduces the cross-section of the shaft at the point where the strain is

transmitted from the spokes to the shaft. The shaft may be used as the windlass direct so that the diameter of the shaft is that of the windlass, or it may be enlarged to form a winding drum of much greater diameter than the shaft. This is determined by the factors of the load and the speed at which this load should be raised.

The man-operated treadwheel crane was used in Roman times and is depicted in the relief in the theatre at Capua. The scale of many Roman buildings and the size of the blocks of stone in them indicate that some form of hoist was in fairly common use. When the great cathedrals were being built in the Romanesque and medieval periods the treadwheel was used in their erection. The length of time required to build the long vaults of the nave and choir would imply that a soundly constructed tread-

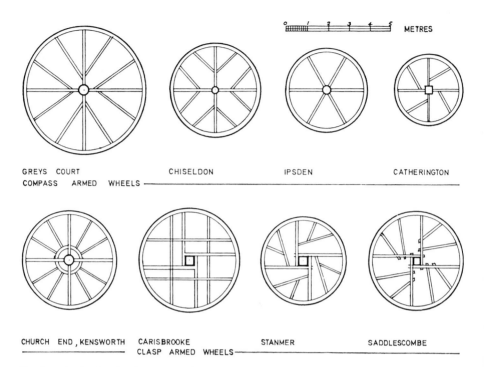

GREYS COURT   CHISELDON   IPSDEN   CATHERINGTON
COMPASS ARMED WHEELS

CHURCH END, KENSWORTH   CARISBROOKE   STANMER   SADDLESCOMBE
CLASP ARMED WHEELS

Donkey wheels, showing the variation between the various types of spokes and the ways the rims are mounted.

wheel was built and fitted to a mobile scaffold so that it could be moved along the roof as each section of the vault was completed. It would also be used to raise the huge roof timbers and the lead for the roof covering. If it survived the arduous duty of hoisting the building materials during the construction then it could be dismantled and rebuilt in the base of the tower above the crossing. From this point materials could be taken to every part of the upper area of the cathedral. Tread-wheel material hoists of this type exist in Canterbury Cathedral and Beverley Minster, whilst that at Salisbury is an intermediate form. Hoists of a dissimilar type but for the same purpose exist at Peterborough Cathedral and Tewkesbury Abbey. In this type there is only a single rim on a single set of spokes, and on this rim there is a series of rungs projecting equally on either side of the rim. The load is raised by the operative climbing hand over hand up the wheel so that his weight causes it to rotate, thereby bringing up the load.

Dockyards needed some sort of mechanical contrivance for lifting loads on to vessels, stepping masts and lifting the great guns into place. Only two dock and wharf cranes remain in Britain: those at Harwich and Guildford. The crane at Harwich was erected in the naval dockyard in 1667 and is a double treadwheel crane, that is a crane in which the windlass portion of the shaft lies between two treadwheels which are fixed in a timber house. At Harwich only the boom of the crane rotated and not the whole crane including the treadwheels, as is usual in continental examples. There were other dockyard cranes, such as Padmore's Great Crane in Bristol and that illustrated by Samuel Scott in his painting 'A Quay on the Thames' of about 1756, which is in the Victoria and

Albert Museum. These appear to have been the type in which the treadwheel was in a fixed housing and only the boom swung to pick up and deposit the load. Both types appeared on the European mainland. The continental examples are all much the same size and the Harwich crane is of a similar size. It is likely that the naval officers and staff returning from Holland with Charles II brought the knowledge with them. The crane at Guildford and one now missing at Stonebridge Wharf on the Wey Navigation are single treadwheel examples, but again with the treadwheels mounted in fixed housings and with rotating booms.

The most numerous examples of treadwheel lifting devices which survive are those associated with water raising. The best known of these is that at Carisbrooke Castle (Isle of Wight), which is made to work for visitors, with a donkey being put into the wheel. This 15 foot 6 inch (4.7 m) diameter wheel was erected in 1587 and the cost of the joinery for it was then £16. The wheel at Greys Court, Rotherfield Greys (Ox-fordshire), is the biggest, being 19 feet (5.8 m) in diameter and 3 feet 10 inches (1.2 m) wide. This wheel was built in a special house in the late sixteenth century over a 200 foot (61m) deep well which had been dug in the twelfth century. The windlass on the shaft had been enlarged and the wheel was worked on a one-up one-down basis.

As with the treadwheel cranes, the wheels vary in their construction as well as their size. There are two types of construction: the compass-armed and the clasp-armed. The compass-armed spokes cannot easily support the rims of these large-diameter wheels without the presence of additional spokes. At Greys Court each principal spoke carries two supplementary spokes so that the rim is supported equally on twelve spokes. At Catherington (Hampshire), the wheel, which is now in the Weald and Downland Open Air Museum at Singleton (West Sussex), was supported on one set of supplementary spokes so that the rim was supported equally at eight points. With clasp-armed wheels the spokes form a

The treadwheel crane at Harwich shown without its building. In this crane the boom rotates and not the treadwheels.

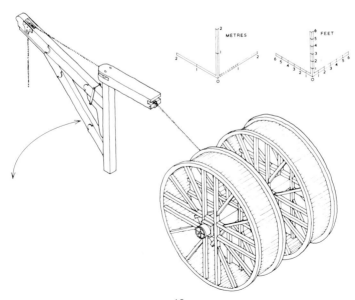

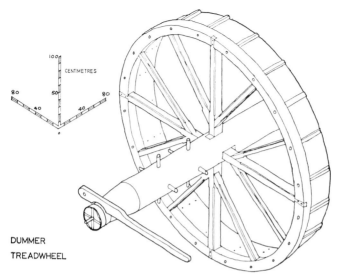

100
CENTIMETRES
80   80
40   40
0

DUMMER

TREADWHEEL

box around a square block on the shaft and are set 'swastika' fashion so that the four main spokes still give equal spacing around the rim. The supplementary spokes are then set on one side of the main spoke to subdivide the spaces on the rim equally.

Some vertical wheels were frequently used only by men, either because their size precluded the use of donkeys or because their situation prevented it. It would not be possible, for example, to introduce a donkey to the wheel inside the Fox and Hounds inn at Beauworth in Hampshire. The wheels at Dummer and Upham, which have only one set of spokes, have too narrow a treadboard on the rim for a donkey to be used.

The vertical hoisting wheel has the greatest significance in the continuity of a technological form for it dates from Roman times to the twentieth century, and it is fortunate that several examples still exist in Britain.

A two-horse winding engine for a colliery. This is from Pyne's 'Microcosm' of 1803.

14

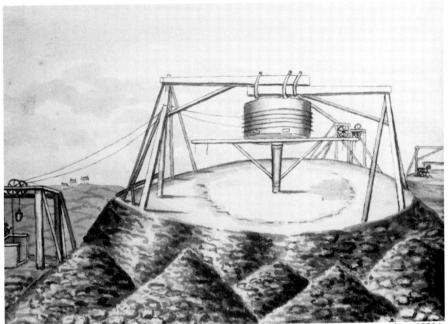

A naive sepia-colour drawing 'Whim for raising ore Gwennap, Cornwall'. Dating from about 1830 it shows a one-up, one-down, two-horse engine for raising ore from the shaft on the left.

HORSE ENGINES IN THE MINES

Early miners could work only as deep as their primitive ladders and crude windlasses would reach. The miners descended by the ladders and sometimes carried the ore back up the ladders, or it was wound out of the mine in buckets or baskets. As mines were sunk to greater depths and because they had to follow the mineral veins, the hoisting of men, ores and surplus water had to be made more efficient, with greater loads carried per journey. The horse engine for direct haulage was used in mines in Europe from the middle ages onwards.

The mining horse engine consists of a heavy duty vertical shaft around which a wide large-diameter winding drum has been built. The drum is at the top of the shaft and below this the horse arm is mounted so that one or more horses can be harnessed to work the engine. A rope runs from the winding drum over the horses' path to a pulley over the centre of the shaft. At Wollaton Hall, Nottingham, there is a preserved mine engine from

Pinxton colliery which was built at Langton colliery in 1844. This example seems typical of most colliery winding horse engines and consists of three elements: the frame supporting the pulley over the shaft; the frame supporting the horse engine; and the winding drum with the horse arm. The winding drum is 9 feet 6 inches (2.9 m) in diameter and 21 inches (530 mm) deep. The horse arm, which is mounted under this, is 13 feet (3.7 m) long to the centre point of the harness frame. The horse arm extends in the opposite direction from the harness frame and drags a sprag around with it. When the horse stopped the sprag dug into the ground and prevented the load in the shaft from running away. If 2½ miles per hour (4 km/h) is taken as the standard speed for a working horse, the horse would give 2.68 revolutions to the winding drum per minute. At that speed the horse would be able to lift a load out of the mine at a rate of 80 feet (24.3 m) per minute. The horse engine at Wollaton

15

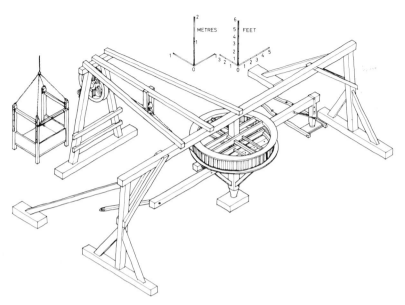

The horse gin which is on display at Wollaton Hall, Nottingham. This was originally built at Langton Colliery in 1844.

Hall ended its days as an engine for shaft inspection; the slow speed and the ability to stop and start were extremely useful attributes.

The most important document associated with mineral working at the beginning of the Renaissance period is Agricola's *De Re Metallica* of 1556, which is a summary of practice at that time. Many examples of the use of animal-powered machines are described, representing practice current at that period and for a long time afterwards. However, medieval European art contains a large number of items showing that the practices illustrated in *De Re Metallica* were universal. The fifteenth-century Kutna Hora *Gradual* shows a conical-roofed horse-engine house in which a horse engine turns a windlass by means of gears. Some examples show a one-up one-down arrangement, as on the Low Hall token, while others, such as that at Wollaton Hall, have only a single rope. These machines have not survived in Britain because they were seldom protected by buildings, but

they had a wide distribution in the eighteenth and early nineteenth centuries. In 1777 the valuation of the gins (horse engines) at Wylam Colliery (Northumberland) read as follows:

Ann Pit 'Whin Gyn 18 ft
  diameter with mettle pully
  wheels'   £20.00
Primrose Pit 'a Whin Gyn
  13 ft diameter with Mettle
  Pullys'   £12.12.0,
  '1 pair swingle trees'   6.6
Wheal Pit 'An old Cogg Gyn
  with screw bolts and good
  iron gear'   £10.10.0
Lane Pit 'A Whin Gyn with
  Iron Wheels'   £30.0.0
Delight Pit 'A Whin Gyn 24 ft
  Diam. no wheels'   £21.0.0

By 1827 there was no gin in use at Wylam, but the Boulton and Watt engine which replaced them was valued at £3500 with all its pumps, boilers and buildings.

The horse engine was often used in the preparation of mine shafts before the

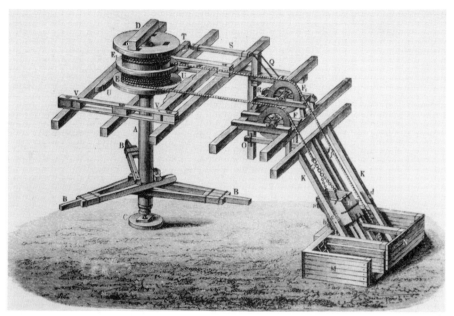

ABOVE: *A three-horse engine for winding coal from a drift mine on the one-up, one-down principle. From Weisbach and Hermann's 'Mechanics of Hoisting Machinery'.*
BELOW: *The mine gin at East Herrington near Sunderland. This winding gear was used to dig the mine shaft and was then retained to lift the pumps and pump rods if repairs had to be carried out.*

head gear and engines were installed, and it was used for the same reason in the creation of shafts and tunnels for canals and railways. Careful search of the ground at the head of tunnel-vent shafts or by mine shafts can often identify the mark of a horse path and the position of the centre of a horse gin. One example can be seen below Wetherlam in Cumbria.

Animal-powered machines have been used in various forms for the dressing of minerals. One form is the edge-runner, in which a roller is rolled over the ore, which has been placed on a circular iron track for crushing, so that the ore particles can be separated from the useless stone surrounding them. In another form, a mass of ore and rubbish was separated by a series of stone blocks dragged around a stone bed by horses. This form is also used to separate silver by creating an amalgam with quicksilver before it is refined.

It is a pity that so few animal-powered machines remain to show what the mining industry was like before the universal use of the steam engine.

THE INDUSTRIAL USES
OF ANIMAL-POWERED MACHINES

There are scarcely any tangible remains of the use of the horse engine as the prime mover in industrial processes. Its use in cotton mills, breweries, foundries and corn mills is documented, but these uses were abandoned before the nineteenth century and have left little trace. The horse-driven corn mill at Woolley Park (Berkshire) still exists on a large private estate and gives an idea of what the machinery was like and how effective it could be. This machine was built in an earlier barn and consists of the horse wheel and the transmission on the ground floor, and on the first floor a hurst frame with two pairs of millstones and a wire brushing machine for dressing the meal. The two horses were harnessed to horns mounted below the 18 foot (5.5 m) diameter wooden gear wheel. The horses were connected to each other by pulleys so that they could pull together. The spur gear of the big wheel engaged with the gear wheel on an upright shaft which extends into the hurst frame above, where by a further gear it drove each of

A direct-drive roller crusher to crush gorse for fodder at Gartly in Scotland. The horse would have been harnessed to the end of the pole with a rotating hook.

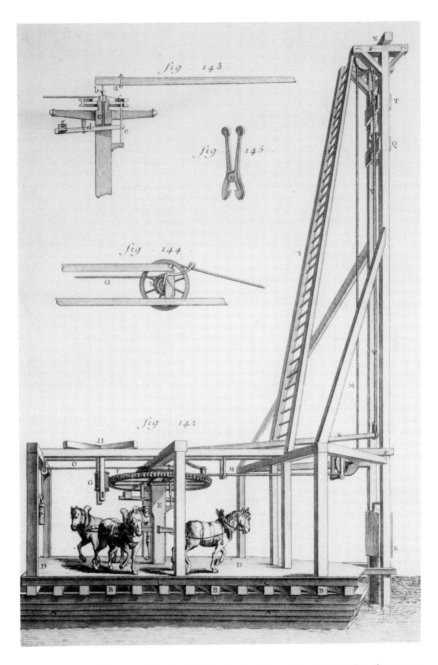

An encyclopedia illustration showing Vaulone's pile driver. Also visible is the release gear, the hammer head and the supplementary windlass on the roof of the horse engine space.

19

The horse-driven corn mill at Woolley Park in Berkshire. This was worked by two horses and could drive two pairs of millstones and a dressing machine for the meal. It is unlikely that more than one pair of stones or the dressing machine were used at one time. Upper, plan of the first floor; lower, plan of the ground floor.

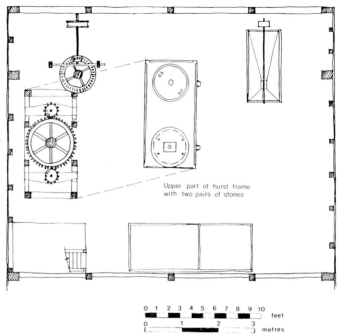

Upper part of hurst frame with two pairs of stones

0 1 2 3 4 5 6 7 8 9 10 feet
0 1 2 3 metres

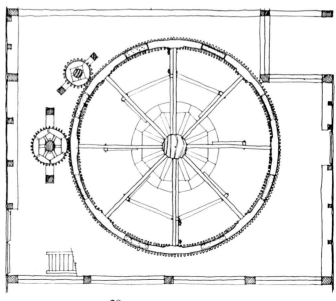

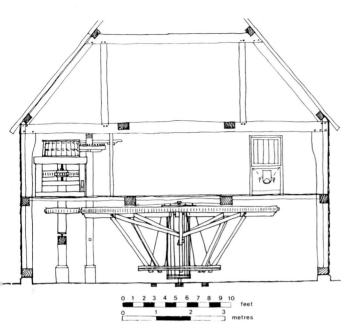

The horse-driven corn mill at Woolley Park in Berkshire. These diagrams show how the drive was taken from the horsewheel on the lower floor to the millstones and dressing machine. Upper, a cross section of the mill; lower, a section showing the horse collars and mill machinery above.

0 1 2 3 4 5 6 7 8 9 10 feet

0 1 2 3 metres

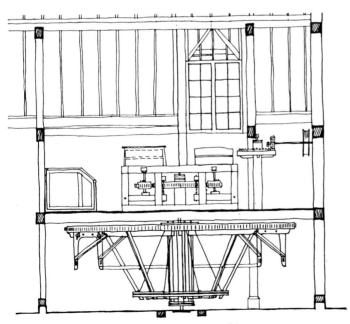

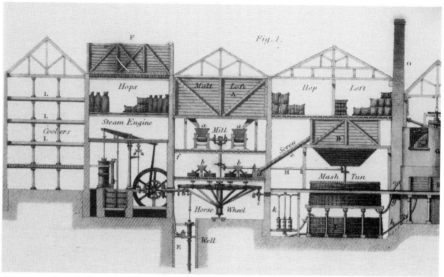

A cross section of a brewery, believed to be the Chiswell Street Brewery, from Rees's 'Cyclopedia'. In this the horse wheel is set between the steam engine, the malt mill and the pumps.

the runner stones. A second upright shaft provided the drive for a pulley system on the wire brushing machine. At 2½ miles per hour (4 km/h) the main gear wheel rotated at 3.8 revolutions per minute and the gear train increased the speed so that the millstones rotated at 92 revolutions per minute. Not many of these machines were erected. J. C. Loudon recorded one he created at Great Tew (Oxfordshire), and another existed at Woburn (Bedfordshire).

In brewing there are well documented examples of horse wheels in use at the breweries at Oxford and Norwich, in Chiswell Street, London, and at the Weevil Brewery, Southampton. An illustration of the Chiswell Street Brewery in Rees's *Cyclopedia* shows the horse wheel as an intermediate wheel inserted between the steam engine and the malt mill in the brewery. If the engine broke down, then the brewery horses could be brought in to drive the mill. The Weevil horse engine was a pumping engine for the brewery erected by John Smeaton in 1780. A low-level crown wheel was mounted below the arms for the two horses, and the drive to the crank and the

pump beam passed below the horse path. As was usual with Smeaton's work, this was an all wooden machine in which the main gear was 19 feet 8 inches (6 m) in diameter and the horses walked in a path 31 feet 9 inches (9.7 m) in diameter. These machines had to be well made in order to give a continuous service and so they could not be of temporary construction.

When Richard Arkwright first started the factory production of cotton in Derbyshire it was in a building powered by a horse wheel, and only when the success of the enterprise outgrew this limited amount of power did he move up the Derwent valley to use the vast amounts of available water power. Unfortunately nothing exists to show how his machines were connected to the horse wheel. Encyclopaedic sources of the second half of the eighteenth century show how the horse wheel was used to power tanning and fulling processes in which a stamping motion was required. Until the Royal Mint became steam powered, its motive power was provided by horse engines.

One industrial use for the horse engine which began in the nineteenth century

22

ABOVE: *A horse-driven apple crusher attached to a cider mill at Foale Leigh in Devon, photographed about 1950. There is no longer a roller crusher in the trough but a pair of rollers in the box on the right.*

BELOW: *An encyclopedia illustration of a machine for fulling cloth.*

23

was wood sawing. The circular saw came into general use at the beginning of the nineteenth century and where it was installed on a permanent site it was powered by steam or water. However, there were then no small portable steam engines in use in the woods, so the horse engine in its portable form was used to power the saws. The cast iron agricultural type of horse wheel was the adaptable form of the engine. The drive shaft, together with an intermediate reduction gear, connected the engine to the circular saw. Another form of horse engine to drive circular saws, which was not very common in Britain, was the paddle engine, or horse treadmill, coupled to the saw by belting and pulleys. In this the horse stood on a belt which moved out from under it and gave power to the shaft at the head of the machine and thence to the pulleys and saws.

It is unfortunate that no examples of the horse wheel and horse engine at work for industry exist, for they would form a link in the knowledge of the power sources in use between the medieval period and the industrial revolution.

ABOVE: *A horse mill at Weyland, Tedburn St Mary, Devon. This photograph is of a wooden horse engine with a cast iron gear ring and dates from the 1920s.*

BELOW: *A permanent building with its horse engine at Scorlinch, Clyst St Lawrence, Devon. This type of building attached to a barn is more commonly found in the north of England and in Scotland.*

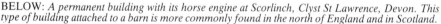

24

A four-horse engine by Barrett, Exall and Andrewes of Reading. This is a 'safety' gear in which all the gears are enclosed in the cylinder.

THE AGRICULTURAL HORSE ENGINE

The horse engine was probably used more in agriculture than in any other industry. In the first instance there were machines which were installed in buildings and which drove threshing machines in adjoining barns. The use of the flail as a means of threshing grain continued in Britain until the nineteenth century and did not end until 1850, when the portable threshing machine had wholly removed the need for hand labour. Andrew Meikle's patent of 1788 marks the time when the thresher was being installed in barns, particularly in the north of England and in Scotland. These threshing machines needed horse engines to power them. The horse-wheel houses, or gingangs as they are sometimes known, were buildings on the sides of threshing barns and were characteristically circular or polygonal. The roof structure would carry a bearing at its centre point which would support the upright shaft of the horse wheel. The shaft would carry a large gear wheel, made of wood, which would be between 10 feet (3 m) and 20 feet (6 m) in diameter. Below this gear wheel two, three, four or six horses

would be harnessed. The drive into the barn would be by means of a lay shaft driven from the gear wheel. Because of the size of the primary gear wheel, the slow speed of the horse — 2½ miles per hour (4 km/h) — could be translated into the necessary high speed of the threshing drums. Because the horse wheel was so efficient, after its introduction for threshing it was also used to drive other fixed machines in barns: water pumps, chaff cutters, turnip choppers and even circular saws. In this way the capitally expensive horse wheel could be put to work throughout the year and not only at harvest time.

When fixed threshing machines were first powered by horse wheels these wheels were made of wood. Cast iron became part of normal millwrighting practice about 1800 and its introduction meant that the large high-level gear of the horse wheel was no longer practical. The result of the change to cast iron was the production of the low-level horse gear. In this the horse or horses were harnessed to a long pole which was mounted on the top of a short vertical shaft. Below this

25

UPPER: *An eight-horse engine on the Isle of Burray, Orkney, in 1908. This is presumed to be in use to drive a threshing machine.*

CENTRE: *A horse engine on a farm near Lustleigh in Devon. This is a portable example, the wheels on the carriage being removable.*

LOWER: *A three-horse engine by Hunts of Earls Colne, Essex, photographed outside the works for their catalogue.*

shaft a first gear wheel, of not more than 5 feet (1.5 m) in diameter, was mounted. This in turn engaged with a bevel wheel and drive shaft. There could be a train of gears on the machine itself or on a frame outside the horse path, to give a higher speed. The low-level horse gear could be fixed in a horse-wheel house or outside a barn, or it could be mounted on a carriage and taken round the farm. The drive shaft was arranged to run along the ground and the horse was trained to step over it.

Several forms of 'safety' horse gear have been patented in which all the moving parts are encased so that animals and operators do not get trapped in the gears. One form, made by the Reading Ironworks, consists of a cast iron cylinder with the horse arm mounted at the top. The main gear wheel is just below the top of the cylinder and the train of gears is below this, with the drive shaft right at the bottom. Wilder's of Reading made a 'safety' gear in which the train of gears is encased in a cast iron dome. These small compact forms were exported widely and both these types have been found in South Africa and in Australia. The low-level horse gear was probably designed for farm use but it was also used for water raising and for circular saw benches, and there are instances of its use for temporary mine haulage and for quarry inclines.

ABOVE: *An unnamed example of a low-level horse engine which was used for pumping water for the 'Roman' bath at Painshill Park, Surrey.*

BELOW: *A safety gear built by the Reading Ironworks Ltd, preserved at the Welsh Folk Museum, St Fagans.*

PLACES TO VISIT

This chapter lists the various sites where animal-powered machines can be seen and gives details of the machines. Readers are advised to check the times and dates of opening before making a special journey to see an animal-powered machine.

Abergavenny Museum, The Castle, Castle Street, Abergavenny, Gwent NP7 5EE. Telephone: Abergavenny (0873) 4282. This museum contains a dog-driven spit. Mounted above this is a stuffed Turnspit, a breed of dog described by Thomas Bewick in his *Quadrupeds*.

Acton Scott Working Farm Museum, Wenlock Lodge, Acton Scott, Church Stretton, Shropshire SY6 6QN. Telephone: Marshbrook (069 46) 306 or 307. This agricultural museum contains exhibits which have particular relevance to the horse. These are: low-level horse engines by Reading Ironworks and Larkworthy of Worcester; a two-horse engine from Penrherber, Newcastle Emlyn (Dyfed); and the horse-driven pumping engine from Swallowfield (Berkshire).

*Ashridge,*Berkhamsted, Hertfordshire HP4 1NS. Telephone: Little Gaddesden (044 284) 3491. There is a donkey wheel at the well house. Limited opening.

Beamish, North of England Open Air Museum, Beamish, Stanley, County Durham DH9 0RG. Telephone: Stanley (0207) 31811. This museum contains the horse-gin house and horse gin from Berwick Hill (Northumberland). This dates from 1814, and the four horses walked in a circle 22 feet (6.7 m) in diameter, with the iron gear wheel mounted above the horse arms. This was typical of threshing machine installations found on many farms in the north of England.

Beaumaris Gaol, Bunkers Hill, Beaumaris, Gwynedd LL58 8EP. Telephone: Beaumaris (0248) 810921. This preserved gaol contains the only *in situ* treadmill in Britain. It is believed to have been installed in 1829 for the punishment of five prisoners and altered in the 1860s to deal with six. Treadmills appear to have been standardised. The wheel was 5 feet (1.5 m) in diameter and the effective tread was 9 inches (230 mm) high. Originally it raised water to tanks on the roof and when the tanks were full the water ran back into the well. Later the prisoners worked against a patent water brake.

Bersham Industrial Heritage Centre, Bersham, Wrexham, Clwyd LL14 4HT. Telephone: Wrexham (0978) 261529. On the lawn in front of the centre stands a full-size reconstructed horse gin of the type used to lift the cage in local coal mines.

Billing Mill Museum, Billing, Northampton. Telephone: Northampton (0604) 408181. The horse engine from Mill Farm, Eversholt (Bedfordshire), where the author found it on a scrap heap, has been re-erected at this museum. An iron upright shaft carries a 9 foot (2.7 m) long horse arm and a 3 foot 3 inch (990 mm) diameter crown wheel which engaged with a 9 inch (230 mm) diameter bevel wheel. The crown wheel was 7 feet (2.1 m) above the horse's path.

Burton Agnes Hall, Burton Agnes, Driffield, North Humberside YO25 0NB. Telephone: Burton Agnes (026 289) 324. This house was supplied with water raised from a well by means of a 12 foot 8 inch (3.9 m) diameter, 2 foot 6 inch (760 mm) wide treadwheel. Its size indicates that it was operated by only one man.

Carisbrooke Castle, Carisbrooke, Newport, Isle of Wight PO30 1XY. Telephone: Isle of Wight (0983) 523112. The 15 foot (4.7 m) diameter donkey wheel was erected in 1587 over a well which had been dug in 1150. The wheel cost £16 to build in 1587. It is walked by donkeys to display its action.

Chilham Castle, Chilham, Canterbury, Kent CT4 8DB. Telephone: Canterbury (0227) 730319. Behind the castle there is a fine wooden horse wheel in a shed. The horse wheel drove pumps by means of a 6 foot (1.8 m) diameter gear wheel mounted on the horse arms.

Dairyland Cornish Country Life Museum, Tresillian Barton, Summercourt, Newquay, Cornwall TR8 5AA. Telephone: Mitchell (087 251) 246. The museum contains an excellent example of a wooden horse gear. It used to drive a barn thresher.

Dummer, Basingstoke, Hampshire. National grid reference: SU 589461. This village was supplied with water by the single-spoked vertical wheel, which is 10 feet (3 m) in diameter. This is an example of a community-owned well and treadwheel and dates from 1879.

Folk Museum of West Yorkshire, Shibden Hall, Halifax, West Yorkshire, HX3 6XG. Telephone: Halifax (0422) 52246. This museum has the horse wheel from a farm at Stillington near York. Three horses were harnessed below a 10 foot (3 m) diameter wooden crown wheel and they walked in a circle 20 feet (6 m) in diameter.

Fox and Hounds Inn, Beauworth, Alresford, Hampshire SO24 0PB. This public house has a 12 foot (3.6 m) diameter man-operated treadwheel in a room behind the bar. It was meant to be operated by one man raising a single bucket.

Greys Court, Rotherfield Greys, Henley-on-Thames, Oxfordshire RG9 4PG. Telephone: Rotherfield Greys (049 17) 529. Water was raised manually from the thirteenth-century well until the well house and donkey wheel were built in the late sixteenth century. The wheel is 19 feet (5.8 m) in diameter and worked two buckets. A horse wheel of about 1870 was brought from Shabden Park (Surrey) and re-erected by the Berkshire Industrial Archaeology Group in 1974. The wheel and pumps are covered by an octagonal roof on cast iron columns.

Guernsey Folk Museum, Saumarez Park, Catel, Guernsey, Channel Islands. Telephone: Guernsey (0481) 55384. The cider barn contains an apple crusher driven by a horse mill.

Guildford, Surrey. Treadwheel Crane. When the centre of Guildford was redeveloped, the wharf and offices of the River Wey Navigation were demolished. The treadwheel crane was re-erected in its building on a new site by the river. The treadwheel is 19 feet (5.8 m) in diameter and 4 feet 8 inches (1.4 m) wide, so two men could tread it comfortably. It is housed in a square building with the crane boom pivoted off the centre of the building.

Harwich, Essex. The Crane. National grid reference: SM 262325. This stands by the church on the sea front. The crane was erected originally in 1667 in the naval dockyard and was moved to its present site in 1930. It consists of two treadwheels, 16 feet (4.9 m) in diameter and 3 feet 7 inches (1.1 m) wide, mounted on a common shaft inside a permanent building. The crane boom is pivoted well forward of the housing so that the crane could deposit materials on the harbour wall.

Hereford and Worcestershire County Museum, Hartlebury Castle, Hartlebury, Kidderminster, Worcestershire DY11 7XZ. Telephone: Hartlebury (0299) 250416. The museum houses a cider mill from Burlingham. This is the typical circular stone trough with an edge-runner stone rotating in it. The harness frame for the attachment of the horse is in front of the main shaft carrying the edge-runner stone so that this was pulled round the stone trough crushing the cider apples.

Industrial Museum, Courtyard Buildings, Wollaton Park, Nottingham, NG8 2AE. Telephone: Nottingham (0602) 284602. In the courtyard of the Industrial Museum a horse gin has been re-erected. This was used for raising men and materials from Pinxton colliery. It was built originally at Langton colliery in 1844 and ceased to work as an inspector's hoist in 1950. The winding drum is 9 feet 6 inches (2.9 m) in diameter and the horse walked in a circle 26 feet (7.9 m) in diameter. The opposite end of the horse arm trailed a sprag, and when the horse stopped the sprag dug into the ground and stopped the cage descending the mine.

Luton Museum and Art Gallery, Wardown Park, Luton, Bedfordshire LU2 7HA. Telephone: Luton (0582) 36941/2. This museum contains the 15 foot (4.6 m) diameter donkey wheel from Nash Farm, Kensworth. Kensworth had other donkey wheels, the best of which is the one at Church End Farm, which is dated 1688 and is also 15 feet (4.6 m) in diameter.

Maidstone Museum and Art Gallery, St Faith's Street, Maidstone, Kent ME14 1LH. Telephone: Maidstone (0622) 54497. This museum has the 12 foot (3.6 m) diameter treadwheel brought from Great Culand Farm, Burham. The size of the wheel would indicate that it was used only by men walking the wheel, but it is known that at the end of its working life it was dog-powered.

Museum of East Anglian Life, Stowmarket, Suffolk 1P14 1DL. Telephone: Stowmarket (0449) 612229. The horse wheel from Henham Hall awaits re-erection. There is a conventional low-level horse gear by Bentall and Company of Heybridge and a horse-driven deep ploughing engine. In this a winding drum hauls a mole plough across a field. By having a horse walking in a circle of large radius, it has the advantage of a lever arm in hauling the heavy plough through the ground.

Museum of English Rural Life, The University, Whiteknights, Reading, Berkshire RG6 2AG. Telephone: Reading (0734) 875123, extension 475. On display there is a horse engine made by Messrs R. Hunt of Earls Colne (Essex), dated about 1910, which drove farm machinery through an intermediate gear, and also a cider mill with a 4 foot 8 inch (1.4 m) diameter edge-runner stone and a 12 foot (3.6 m) diameter stone crushing trough. The library and archives contain many photographs and catalogues of animal-powered machines.

Museum of Lakeland Life and Industry, Abbot Hall, Kendal, Cumbria LA9 5AL. Telephone: Kendal (0539) 22464. A horse gear from a farm at Cartmel in Furness has been re-erected here. This is a low-level horse engine with a main gear wheel 4 feet (1.2 m) in diameter. Although the horse arm is missing, it was intended to be powered by two horses. It was made by J. Scott of Belfast.

Odin Mine, Castleton, Derbyshire. National grid reference: SK 135835. On the open fell near mine shafts can be seen the derelict remains of an ore crusher. This consisted of a cast iron crushing ring track 18 feet (5.5 m) in diameter, on which a 5 foot 3 inch (1.6 m) diameter, 12 inch (300 mm) wide gritstone crusher, bound with a 12 inch (300 mm) wide iron tyre, was rotated by a horse. The horse arm passed from the centre of the crushing ring through a hole in the centre of the gritstone crusher. The ore was shovelled on to the iron track and was crushed by the passage of the heavy roller crusher before the waste was washed out.

Ryedale Folk Museum, Hutton-le-Hole, York YO6 6UA. Telephone: Lastingham (075 15) 367. This museum has a low-level horse engine mounted on a timber chassis. The shaft carries a canister in which three horse arms were bolted.

Salisbury Cathedral, Salisbury, Wiltshire. A fine treadwheel may be seen by visitors to the tower. The 10 foot 10 inch (3.3 m) diameter wheel was trodden by two men on either side of the single ring of spokes, and it raised material from the floor of the cathedral through the eye of the crossing vault.

Science Museum, Exhibition Road, South Kensington, London SW7 2DD. Telephone: 01-589 3456. A full-sized horse wheel and churn from Broughton Farm, near Aylesbury, have been set up here. A horse, walking under the 15 foot (4.6 m) gear wheel, drove the butter churn by means of a vertical shaft. Among the models there is a horse-driven rag-and-chain pump, a Middle Eastern chain of pots driven by an ox or a camel, and trade models of a Larkworthy horse engine and an inclined plane horse gear.

Sutton, Thirsk, North Yorkshire. On the village green there is a preserved crab (gorse crushing) mill. This mill is an edge-runner stone which is drawn round by a horse in a circle over a flat base. The edge-runner crushed gorse to make it edible for farm animals.

Weald and Downland Open Air Museum, Singleton, Chichester, West Sussex PO18 0EU. Telephone: Singleton (024 363) 348. There are three animal-powered machines on the museum site: an early seventeenth-century treadwheel for raising water which came from Catherington (Hampshire); a cast iron horse engine by Warner of Cricklewood, which supplied water to a house in Patching (West Sussex); and in a hexagonal brick building a horse-driven pug mill, which came from a brickworks at Redford (West Sussex).

Welsh Folk Museum, St Fagans, Cardiff, South Glamorgan CF5 6XB. Telephone: Cardiff (0222) 569441. There are several examples of animal-powered machines in this museum. In the kitchen of the manor house there is a dog-driven spit. The wooden treadwheel, or dog cage, is 4 feet 8 inches (1.4 m) in diameter. In the museum block there is an oblique dog-driven wheel, which came from Dulellog, Rhostryfan, Caernarfon. The wheel is 9 feet (2.7 m) in diameter and drove a churn from a gear under the boards of the wheel. In front of the barn by the long ponds there is a horse engine which came from a farm in the Gower; this is a Reading Ironworks 'safety gear'. The gear train is encased in a cast iron cylinder and the drive runs into the adjacent barn at ground level.

BIBLIOGRAPHY

Agricola, Georgius (Georg Bauer). *De Re Metallica*, Basle, 1556. (Translated by Hoover, Herbert C., 1912.)

Atkinson, Frank. *The Great Northern Coalfield 1700-1900.* Durham County Historical Society, 1966.

Atkinson, Frank. 'The Horse as a Source of Rotary Power'. *Transactions Newcomen Society*, Volume 33, 1960-1.

Bennett, R., and Elton, J. *History of Corn Milling,* Volume I (Handstones, Slave and Cattle Mills). Simpkin, Marshall, 1898.

Brunner, Hugo. 'Relics of the Wheelwright's Craft, Donkey Wheels as a Source of Power'. *Country Life*, 28th December 1972.

Brunner, Hugo, and Major, J. Kenneth. *Water Raising by Animal Power.* Privately published, 1972. (An amplification of papers published in *Industrial Archaeology* Volume 9, Number 2, 1972.)

Clark, Duncan W. 'The Harwich Crane'. *Essex Review* 165, Volume XVII, January 1933.

Clark, H. O. 'Notes on Horse Mills'. *Transactions Newcomen Society,* Volume 8, 1927.

Corley, T. A. B. 'Barratt, Exall and Andrewes, The Partnership Era 1818-1864'. *Berkshire Archaeological Society Journal* Volume 67, 1975.

D'Acres, R. *The Art of Water-Drawing.* London 1660. (Reprinted by The Newcomen Society, 1930.)

Davies-Shiel, M., and Marshall, J. D. *Industrial Archaeology of the Lake Counties.* David and Charles, 1969.

Dickinson, H. W. *Water Supply of Greater London.* The Newcomen Society, 1954.

Drachmann, A. G. *Ancient Oil Mills and Presses.* Copenhagen, 1932.

Ewbank, T. *Hydraulics and Mechanics.* David Bogue, 1859.

Gille, B. *The Renaissance Engineers.* Lund Humphries, 1966.

Harrison, A. and J. K. 'The Horse Wheel in North Yorkshire'. *Industrial Archaeology* Volume 10, 1973.

Hewitt, Cecil. *English Cathedral Carpentry.* Wayland, 1974.

Higgs, John. *The Land.* Studio Vista, 1964.

Hodges, Henry. *Technology in the Ancient World.* Penguin, 1971.

Hook, B., and Murlees, B. J. 'Horse Gins in Somerset'. *Somerset Industrial Archaeology Society Journal* Number 1, 1975.

Hutton, Kenneth. 'The Distribution of Wheelhouses in the British Isles'. *The Agricultural History Review,* Volume 24, Part 1, 1976.

Jespersen, Anders (editor). *Report on Watermills,* Volume 3. Copenhagen, 1957.

Jewell, C. A. *Victorian Farming.* Barry Shurlock, 1975.

Keller, A. G. 'The Oblique Treadmill of the Renaissance: Theory and Reality'. *Transactions of the Second International Symposium on Molinology.* Denmark, 1969.

Kretzschmer, Fritz. *Bildokumente Römischer Technik.* Vereins Deutscher Ingenieure, 1978.

Loudon, J. C. *An Encyclopedia of Agriculture.* Longman, 1844.

Major, J. Kenneth. *Animal-Powered Engines.* Batsford, 1978.

Major, J. Kenneth. 'The Horse-Driven Corn Mill in England'. *Transactions of the Second International Symposium on Molinology*. Denmark, 1969.
Monnet, A. G. *A Treatise on the Exploitation of Mines*. Paris, 1773. (Reprinted by Clough, 1974.)
Moritz, L. A. *Grain Mills and Flour in Classical Antiquity*. Oxford University Press, 1958.
Natrus, L. van, Polly, J., and Vuuren, C. van. *Groot Volkomen Molenboek*. Amsterdam, 1734. (Reprinted by Kampen and Zoon, 1969.)
Nixon, Frank. *Industrial Archaeology of Derbyshire*. David and Charles, 1969.
Partridge, M. *Farming Tools through the Ages*. Osprey, 1973.
Pyne, William H. *Microcosm*. London, 1806. (Reprinted by Luton Museums, 1974.)
Ramelli, Agostino. *Le Diverse et Artificiose Machine*. Paris, 1588. (Reprinted by Gregg International, 1970.)
Rankine, W. J. M. *A Manual of the Steam Engine and Other Prime Movers*. Charles Griffin, 1859.
Riley, A. J. 'When Horses Turned the Mill'. *Country Life*, 27th November 1972.
Salzman, L. F. *Building in England down to 1540*. Oxford University Press, 1952.
Schiøler, Thorkild. *Roman and Islamic Water-Lifting Wheels*. Odense University Press, 1973.
Steele-Elliott, J. 'Bygone Water Supplies'. The Bedfordshire Historical Record Society, *Survey of Ancient Buildings II*, 1933.
Tomlinson, C. *Illustrations of Useful Arts and Manufactures*. George Virtue, 1854.
Walton, James. *Watermills, Windmills and Horse Mills of South Africa*. C. Struik, 1974.
The Catalogue Collection of the Museum of English Rural Life, Reading.

A locomotive driven by an oblique treadmill and four horses. This was invented in Italy in 1850 and shown on the London and South Western Railway.